The Write Equation

Writing in the Mathematics Classroom

- John A. Carter
- Dorothy E. Carter

Dale Seymour Publications

Managing Editor: Mike Kane
Project Editor: Joan Gideon
Production Coordinator: Karen Edmonds
Cover Art: Rachel Gage
Designer: Detta Penna

This book is published by Dale Seymour Publications, an imprint of Addison-Wesley's Alternative Publishing Group.

Order Number DS21214

ISBN 0-86651-671-9

5 6 7 8 9 10-ML-97

Contents

Foreword

As teachers, we are always seeking new ways of explaining mathematical ideas and concepts, hoping to improve our students' understanding. Often one or more of the following approaches may be used: the visual, the analytic, the algebraic, or the geometric. Each of these needs to be reinforced both orally and in written form on an almost daily basis.

The power of words is a phenomenon of human experience. The ability to use this power opens new doors. Without it, limitations may be placed on our ability to grow and develop.

As a boy growing up on a dairy farm in Illinois, I was encouraged to do the very best I could in school—I had the parental support that we so much desire for all our students. However, my schooling experience demanded very little when it came to expressing my thoughts and ideas, either verbally or in written form. I often found myself wondering if my ideas ever really mattered to my teachers. The only exception came in writing proofs in plane and solid geometry. As a result, I had great doubts about my ability to write. I avoided courses in college that required writing, even though I had the God-given gift, unknown to me at the time, which was yet to be exercised and developed. Not until I was doing graduate work at the age of 24 did writing become commonplace in my academic life. How sad that was, as I think back on it now!

As mathematics teachers, we owe it to our students to value their ideas and thoughts as well as to help them develop mathematically. We must require our students to express themselves both orally and in written form on an almost daily basis.

You are about to read a book that will provide you with a rich set of exciting ideas, all of which are designed to assist your students in developing their abilities to analyze, communicate, and value their mathematical thoughts and ideas.

Lee E. Yunker

Chapter One

Why Write in the Mathematics Classroom?

Recently while walking down the school hallway between periods, one student was heard complaining to another, "Why do we have to write in math class? When will I ever need to write about math in the real world?" Surprisingly, the other student remarked, "Are you kidding? The real world is all about writing! You'll have to write reports and share evidence—you'll write about math all the time!" These types of complaints are heard frequently around our schools, and it is refreshing to know some students are finally beginning to understand the reason for writing in the mathematics classrooms. We look forward to the day when students ask, "But why can't I write my answer in a composition?" However, until that time, we must be ready to address the issue of why writing is so important in mathematics.

A call to reform

As teachers in mathematics classrooms begin to use writing as a tool for communication, assessment, and expression of ideas, they will undoubtedly be asked to answer a variety of questions posed by students, parents, colleagues, administrators, and other concerned people. Students may be heard saying, "Why do we have to write in math class?" Parents often have mixed emotions: "Why should my son have to write words in math? When I learned math, it was just numbers!" or "Are you sure this will benefit my child? I remember New Math." Administrators and school board members have concerns about implementing innovative techniques in their districts. Whatever the questions and whoever is doing the asking, it is important that the teacher planning to implement a writing program in the mathematics classroom feel comfortable with the idea. The teacher must be willing to have fun and explore with this tool, which until recently was virtually nonexistent in mathematics classes.

Mathematics teachers across the nation are attempting to answer the call to reform mathematics education. They are trying to give students a feeling for what it means to understand, to use, and to explore mathematics. It is precisely this feeling, along with an understanding of the usefulness of mathematical thought, that many students over the past several decades never had. Recent documents such as the National Council of Teachers of Mathematics' *Curriculum and Evaluation Standards for School Mathematics* and The National Research Council's *Everyone Counts* have brought to our attention the need for changing our attitudes in the teaching of mathematics. The NCTM *Standards* recommend that for mathematics to be taught in an effective and meaningful manner, students must be given an "inclination to monitor and reflect on their own thinking and performance."[1] The National Council of Teachers of English has long advocated using writing to learn, not just learning to write. Both these recommendations present to educators an irresistible device for improving the understanding of mathematics.

An alternative to traditional assessment

Composition is an important tool in mathematics because it allows students to focus their attention on what it is they are doing. They need not be preoccupied with getting the right answer or asking questions like, "How do I do this one?" but rather they can concentrate on the process of solving a problem or the understanding of mathematical concepts. For example, ask any freshman algebra I class the question, "What is the commutative property?" You are sure to get a variety of responses similar to these:

a) $A + B = B + A$

b) $AB = BA$

c) $A + 0 = A$

d) I don't know.

e) It's a property that I'm supposed to know.

None of these responses really gives the reader any idea as to whether the person giving the response understands the commutative property, has some idea about the commutative property, or is totally lost. Alternatively posed, the question can read, "Summarize the main ideas behind the commutative property without the use of mathematical symbols." Now each response reveals a great deal of information about the student's understanding:

a) It means that two things can be done in any order if they're the same thing.

b) It says that two numbers can be added or multiplied in any order.

c) The commutative property is that 2 plus 3 and 3 plus 2 both give you 5 'cause it's addition and 3 times 2 equals 6 and 2 times 3 equals 6 'cause it's times.

d) It says anything plus zero is the same thing.

For the teacher, the information gained from each of these four attempts is priceless. It is far more information than can be obtained using a multiple choice format or recitation of memorized formulas. When writing their responses, students are forced to rely on words to illustrate their understanding rather than a string of symbols that may mean nothing to them. Assessment of students' knowledge is more complete because one can pinpoint errors in the students' conceptual development. When grading a paper with rote, repetitive drill problems on it, the teacher cannot determine whether an error is due to guessing, carelessness, or misunderstanding of the math concept. Or perhaps the error reflects the inability to memorize a property. However, when a student writes an essay summarizing a concept, the teacher can readily determine whether the student has understood the idea.

Of course, writing in the mathematics class has other benefits besides better assessment of mathematics skills. In "A New Way of Thinking: The Challenge of the Future," Sam Crowell declares, "The greatest challenge facing education is . . . the need to discover with our students a new way of thinking."[2] Writing about mathematics helps students see new connections between the various subjects they are studying. Students need to understand that each class does not contain a set of isolated skills, and that writing skills and mathematics skills are not left behind the minute they leave the English or mathematics classroom. In both classrooms, students are able to practice and improve their writing skills and use writing as a problem solving tool and a means of self-expression. Students are then better able to see the connection between mathematics, writing, learning, and the real world. Jean Kerr Stenmark writes, "In the world of work, people are valued for the tasks and projects that they do, their ability to work with others, and their responses to problem situations. To prepare students for future success, both curriculum and assessment must promote this kind of performance."[3] Judah L. Schwartz adds, "If we use tests that ask students to recognize answers rather than construct solutions, we will be teaching students tricks for recognizing answers rather than strategies for constructing solutions. That is what most test preparation is now about. If we use tests that suppress subtlety and nuance, we should not be surprised that our students' analyses tend to be superficial and simplistic."[4]

There is agreement that reform is needed, and therefore we offer one method of change—encourage students to write in the mathematics classroom. Incorporating writing within the mathematics classroom can support current teaching structures and be a valuable alternative to the traditional modes of assessing the understanding of mathematics.

Chapter Two

Starting Your Program

Once a teacher or department has decided to implement a writing program, the real process is just beginning. The planning and questioning that goes into developing the program is crucial. Following are some of the areas that ideally should be considered before the implementation of a writing program. After the program is up and running, discussion will continue in many of these areas. It is likely that as a staff implements a program, the initial decisions with respect to these questions will seem inadequate and will need to be reexamined and changed. It is important that teachers be willing to allow the program to remain dynamic. Insistence that the program be set in stone can curb the creativity of those involved.

When should I start?

Many teachers are comfortable starting the program immediately with the start of the new school year. If this is the students' first experience with writing in mathematics, this can lead some students to form sour opinions the very first day. We recommend that teachers take a few weeks to get to know their students before asking for their first mathematics writing sample. An appropriate first sample may be a journal entry recording their opinions of the class or a comparison between their expectations for the class and their experiences so far. Whatever the first sample is, it may take some prodding and coaching to help first-timers feel at ease with what they view as a non-mathematical assignment.

The grade level at which writing is used should also be chosen carefully. Writing in the middle school or the freshman general mathematics class is likely to take more time for implementation than it would in an advanced pre calculus or trigonometry class. Whichever level is chosen, it is important to take time to explain to students the reasons for asking them to write. Without such an explanation, some students will feel that this is simply a new way to impose more work. This misunderstanding needs to be addressed as soon as it appears. Students and parents who adopt this view of writing will resist efforts to implement the program. With an appropriate introduction and some positive feedback for their attempts, students will be very willing to continue writing throughout the year.

How often should I ask students to write?

Because many good intentions sometimes fall victim to lack of time, it is a good idea to commit yourself to a certain number of writing experiences before the implementation of the program. In the first year of a program, this commitment may range anywhere from once per chapter to a minimum of three times per week. The choice is yours. Be sure to choose a realistic goal. If the goal is unattainable, students will notice the increased absence of writing experiences and assume that you have lost faith in their value. A reasonable goal to start with would be two times per chapter. If you are starting the program as a team of teachers, a goal of four times per chapter is not unreasonable.

What kinds of writing experiences are there?

Writing experiences may be classified into three categories. They are journal writing, essay composition, and research and project writing. Each of these categories is explained in detail in the following chapters, but a brief overview is given here.

Journal Writing

The major focus of journal writing should be the quick assessment of opinions and feelings on an issue. A journal prompt may ask students to write for ten to fifteen minutes.

Essay Composition

An essay is more formal than a journal entry and requires higher-order thinking skills in many different ways. Students must be able to formalize concepts they have learned, organize them, and commit them to paper in an acceptable fashion. Essay topics are generally content-oriented and conceptual in nature. An essay may take one of several different forms. It can be persuasive, descriptive, expository, definitional, or narrative in nature. It can also focus on comparing and contrasting ideas.

Research and Project Writing

The overall focus of this type of writing is to help students make connections to other topics or areas of study. It offers some freedom in terms of topic choices, but it is focused enough to develop the skills and processes desired.

How do I teach writing?

After choosing the type of experience, you must write questions and prompts that elicit the information you want to know. However, simply giving students a prompt and time to write will probably produce poor results. If your results are poor, you may become discouraged with this form of assessment and perhaps

SUMMARY OF WRITING EXPERIENCES

Journal Writing

- Non-threatening
- Personal
- Informal
- Free Writing
- Statement of feelings and opinions
- Not evaluated on content

Essay Writing

- Analytical
- Narrative
- Explanatory
- Coherent
- Focused
- Organized
- Descriptive
- Persuasive
- Unified
- Compare/contrast
- Makes connections

Research and Project Writing

- Self-motivated
- Purposeful
- Real-world applications
- In-depth research
- Interviews and questionnaires
- Extrapolation
- Collaborative (optional)
- Answers "Why are we doing this?"

put it back in the file cabinet. Instead, take some time to explain clearly to students your expectations for their work. To begin, clearly word the writing assignment and state the kind of writing you want from them. For example, if you want students to compare and contrast two types of problem solving strategies, make sure the words *compare* and *contrast* appear in the instructions.

Second, help students develop some strategies for writing. Brainstorm as a class, or encourage students to suggest possible topics for the assignment. Role play the thinking you would do if you were presented with the same writing assignment.

Third, model the kind of response you want by providing an example for students to read. Do not provide an example using the topic you want the students to address or you will have thirty papers that sound vaguely familiar. Choose a different topic. Also, be sure your model provides students with a clear idea of the format you expect. If you want a well-organized essay, do not assume that students know how to write one. Provide some examples for them to use as guidelines. Highlight features that are important to you, such as the thesis sentence or complete sentences.

Finally, you may want to give students an opportunity to share their work with others before they turn it in to you. This provides them with a chance to proofread and edit each other's work as well as to share their ideas with others. Students often learn something new about a topic by reading and criticizing each other's work. You might provide a checklist for students to use when they are peer editing.

How do I evaluate ?

Several techniques will be discussed in detail in chapter six. Currently many teachers use one of the following tools:

Writing Portfolios. Portfolios allow students the chance to collect a representative sampling of their work over a designated period of time.

Checklist. Checklists give students valuable information about the expected quality of their performance and the criteria by which it will be evaluated.

Peer Evaluation. This form of evaluation gives students the opportunity to aid each other in the creative process. Students gain insights from the diversity among their classmates.

Holistic Evaluation. This type of evaluation examines the overall quality of a student's work.

Increasing the chances of success

To increase you chances of successfully implementing the program, determine beforehand your level of commitment to writing in mathematics. Avoid becoming so zealous that you are grading compositions every night. Also avoid making only a half-hearted attempt. Your students will soon sense you are not serious about the writing, and they may adopt your attitude. Instead, you might decide not to grade every assignment that you give, especially journal entries, but rather ask students to turn in their best writing during a given period of time for a grade.

Second, explain the reason for writing and give your expectations and requirements. Yes, this means taking class time. However, rushing through the writing instruction will result in disappointing work, and both you and your students will become discouraged.

Third, consider why you are adding writing to the curriculum and keep that reason in mind as you evaluate the papers. A common temptation is to mark (in red!) every spelling error and comma splice, but is that why you are asking them to write? Allow students to experience some success in this alternative form of assessment so that they develop positive attitudes toward it.

Chapter Three

Journal Writing Experiences

What is a journal?

Journal writing has long been a popular form of self-expression. People as varied as Henry David Thoreau and Anne Frank have used journals as creative and introspective outlets for their thoughts. Journals in the classroom can be used in much the same manner. In English class, journals are kept for students to record their opinions on issues ranging from topics in literature to world events, or to serve as a prewriting activity. Now students are given the same opportunities to explore their opinions on mathematics topics and processes. In both cases, the emphasis is on an honest and thoughtful response to the topic at hand. Students are not necessarily asked to follow a specific organizational format when writing. Journal writing is meant to be a nonthreatening writing opportunity in which students are able to express their opinions, possibly about controversial issues, and evaluate their progress in the class.

Journal writing can be implemented in many ways, but we have found success with two methods. In the first method, students are asked to keep a journal in a section of their mathematics notebooks. Students may write in their journal every day or less often, depending on the teacher's preference. We suggest asking students to use ink for their journal writing so that they will be able to look back on their progress through their journals in later years. Ink also gives the journal a neater appearance. The teacher may or may not provide a writing prompt or might even ask students to suggest journal topics. Usually ten minutes is given to journal writing, and then members of the class are invited to read their entries aloud.

Using the second method of journal writing, the teacher duplicates a specific journal topic for the students to address. After allowing students ten minutes to write, the teacher collects the papers and chooses some to read to the class.

How is it useful?

Journal writing has been successful for several reasons. First, the topics for journal writing are designed so that students are always able to give an answer. Many times the journal topic asks students to give opinions on a topic they have been studying in class. For example, students might be asked to write journal entries on whether they believe factoring is an important skill to know. This provides the students with an opportunity to vent any feelings and frustrations they may have about factoring. A student cannot honestly answer this question by saying, "I don't know," because everyone has an opinion, even if it may simply be "I don't care."

Furthermore, some journal topics allow students to explore connections between mathematics and other subjects. For example, students might be asked to explain what forms of mathematics are used in any other class they are taking. After students have written this journal entry, the teacher might ask them to share their responses with the rest of the class, thereby broadening other students' perspectives on applications for mathematics.

Journals are also effective writing tools because they are nonthreatening. Students commonly worry whether they are writing the right way, or ask, "Is this what you wanted me to write?" This inhibition forces students to ignore their true feelings in order to please the teacher. However, in journal writing we stress that there really is no single right answer or correct way of responding to the issue. We tell students we are interested in their opinions, and we want them to be honest. The few requirements we make in journal writing are that the student address all the issues, and perhaps that the student complete a specified number of journal entries in a given period of time. Students are not graded on content. It is suggested that the teacher initially provide students with a few sample journal entries so students know exactly what is expected of them. We recommend that teachers not become overly critical in evaluating student journals. Journal entries should be thought of as rough drafts—the basis of a more in-depth and thoughtful discussion.

Another benefit of journal writing is that it enables the teacher to become aware of specific concerns or problems students may be having. For example, if students complain in their journals that there seems to be no purpose in studying a certain topic, the teacher might be able to address that issue and therefore make the topic more meaningful to the students. A frequent journal prompt might be "What did you learn today in class?" "How did you do on the homework assignment?" or "What are some of your goals to help make your performance on this chapter an improvement over the last one?" These questions provide students with the opportunity to have a written dialogue with their instructor and do some self-evaluation of their progress in the course.

Because journals are not graded on form they are fairly easy to check frequently. Teachers can focus their time on skimming the entries for ideas. Journals can be used to survey the class on an interesting topic, to brainstorm for a later research project or more formal composition, or to monitor student understanding of a concept quickly and informally. Teachers might collect the journals once a week and write brief responses back to the students. Or teachers might have students read journal entries aloud, thus providing a chance for instant feedback. In either case, teachers can quickly evaluate the journal entries and assign a grade based on the number of entries or the thoughtfulness of the responses.

Sample topics and student responses

Great care should be given to choosing a journal topic so that students have an opportunity to explore their attitudes about the class. Good journal prompts are fairly easy to write, provided they offer students a chance to give personal responses. Most students will not develop their ideas when given simple yes or no questions, so such prompts should be avoided. However, it is not always enough simply to ask why or why not on a question either. For example, students will often take the easy way out with "No, it's boring" when faced with a question like "Do you like mathematics? Why or why not?" Try to be more specific, challenging students' critical thinking abilities with the journal topic. By changing "Do you like mathematics?" to more specific and relevant questions like "Explain a time when mathematics class has challenged or excited you," or "When has mathematics frustrated you?" teachers are able to focus students' thinking and encourage more enlightened responses.

We have found that journals can be effective writing tools in mathematics because of the freedom they allow students in expressing their ideas. We now offer several examples of journal writing topics and student entries. The student entries are not without their flaws in grammar and composition. Nonetheless the students' ideas are worth noting.

Journal Topic: The governor of one of our states has recently made his opinions known about high school mathematics. He feels that in order to receive a diploma in an American high school, a student should have completed four years of high school mathematics. The public universities in Illinois have already said that, starting in the fall of 1993, every student they admit must have completed at least three full years of college preparatory mathematics—algebra I, geometry, and algebra II.

a) What do you think about the universities' requirement for admission?

b) What do you think about the governor's idea?

c) What do you think the requirement should be? Why?

I think you should only have to have 3 because what else do you need to know about mathematics? I think the governor's idea is stupid.

Andre

I think the governor has a good idea about this requirement. Every student should complete four years of math. Well, if a student wants to go to college of course they will meet the requirement. What would a person with less than 3 years of math accomplish in life?

Elisa

I don't really understand what there is to think about the requirements. I will probably take 3 years of math anyway, and if I had to I would take four. To me, math is one of the easiest subjects so I don't care if it is required or not. I don't think there should be any requirements, you should just take what you want.

Javier

I think that it is good that they are making it more difficult to get into college. I don't think that it is necessary although to have to take 4 years of math for those who have no relation to math in their field of interest. I think the students should be required to take math up to geometry. If they start their freshman year at geometry then they don't need any other math unless they want to or their teacher felt it's necessary. I think that the school should enforce math but not demand it.

Sheila

Journal Topic: Over the years many people have argued about how to teach inequalities in algebra I. Some say that it is similar to solving equations. Others argue that there are some new things that students must be aware of when they solve inequalities.

a) Take a position on this topic.

b) Describe the evidence you can use to support your position.

I think it should be taught because they're not always going to be equal and kids have to know that with inequalities. They have to know that if you multiply by a negative in an inequality you have to change the sign in order for it to be true. There are differences like with equalities you do them and check them but with inequalities you do them, check them, and graph them. You have to learn about what way the arrow goes and if it's open or closed.

Keisha

I don't think that solving equations is the same as inequalities because you have to prove it. When you're solving equations you tell what x is equal to, but when you solve inequalities they don't have to be equal—they can be < or >. So you don't really know what x is you just know if it is < or > something.

Rodolfa

I feel that inequalities don't play an important position in our math program. I have seen how business accounts are run and the different fields of jobs that I like and that most other people do too. I've never seen an accountant use those types of formulas showing x is less or greater than. They won't show an "< >" they'll just show the raw # facts. Not guesses, just the numbers. Maybe if you have a loser job you might see this but I am trying for a good job. I like how the equations are solved for the formulas and things that will come in helpful.

Alex

Algebra is sort of neat in a way I've always loved math but I never did my homework though I did understand what we were doing. linequalities are different and should be taught because it's not the same as equalities because you find specific answers in equalities. like I said equalities you find certain specific answers using APE MPE and other rules for algebra when learning inequalities it's good to know equalities before. because they are solved in a similar way the only thing different is that they make you think a little more and when you MPE with negatives you have to change the sign. The End.

Lorenzo

These examples clearly show that students have opinions about what is going on in their mathematics class. Some students addressed the issues better than others, and some used their journals as a sounding board for their frustrations. All of the students were able to express what they knew about the topic and provided the teacher with the opportunity to react to some misconceptions about important concepts. For example, students provided contrasting responses to the first question. The instructor might have Elisa explain her view to Andre in an attempt to address Andre's rather hostile view of mathematics requirements. Alex's comment on the second topic about how only "loser" jobs would use inequalities should be addressed with examples showing how many jobs do require the knowledge of inequalities.

Some other prompts that can be used are:

- What were the goals that Mr. Carter had in mind for today's lesson? What did he do that aided me in learning the material?
- If I could have written the lesson today, I would have done things differently to make sure everyone learned the main ideas. I would have . . .
- If I could have taught the lesson today, I would have . . .
- I was/was not excited about coming into algebra class today because . . .
- I do/do not see how this relates to anything that I already know. Explain your response in detail.
- Three things I learned during class today are . . .
- Today is the first day of a new quarter (semester, chapter, unit). Take this chance to think about what you would like to change about your performance in class last quarter (semester, chapter, unit). In your journal entry for today, list the

behaviors that you would like to change and state some specific goals for yourself. Let Mr. Carter know if there is any way he can help you reach those goals.

Giving students an opportunity for self-evaluation can be very helpful. Although, they are sometimes harsh judges of their own work, their comments can also provide a teacher with insight into how they have improved. The following excerpts from students' writings illustrate this aspect of journal writing.

Journal Topic: Having just completed a fifteen-day unit on quadratics and other polynomials, take some time to think about your performance and understanding throughout the chapter. In a personalized journal entry, write your evaluation of the unit.

In this chapter I learned many new things. It was broken into three major parts: quadratics, polynomials and geometry based problems. I was very interested and amazed at most times in this unit.

Craig

I have a little story to tell you. Well, maybe it's a little longer than little, but here it goes anyway. (Before I get started though, let me tell you a thing or two. If you get bored reading this, you can quit anytime you'd like, just give me a good grade.) This chapter I felt like I was going to die right in the middle of the classroom. Not because I thought it was boring stuff or anything, but because I didn't understand quite a few things. Still, I like challenges though.

Take multiplying trinomials for instance. On the day or days when we were doing those things I kept both my feet and fingers crossed, hoping Mr. Carter doesn't call on me. At times it has worked and there have been times when he has called me on an extremely hard problem. Take the time when he called on me to give an answer to a very hard problem, at least to me it was hard. (Mr. Carter, are you bored yet?) It was when he asked me to do a trinomial problem in which you had to multiply to get the right answer. I was sweating like crazy (at least it seemed that way) and everyone was looking at me. I was thinking "Oh MON DIEU, QUELLE HORREUR!"

Anyways, this chapter had its ups too. It was fun though because I was learning how to sing and do math at the same time. There was a funny time I remember. It was also when he called on me to answer a problem. (Mr. Carter, are you still reading this?) I knew what the answer was but I had kept my mouth shut because I had a humongous wad of gum in my mouth. People must have thought I was a bubblehead, even the teacher. Someone else gave the answer.

In math class I will try my very hardest from now on. I'll try to look over my notes everyday and hope to become smart in the subject. I'll also practice singing at home and I'll try not to ask people for too much gum to avoid future problems. Still, this math class is fun and quite a bit of a challenge but I know I can do it if I try.

(Did you get bored?)

Manuel

I thought this chapter was challenging in a way but also quite easy. I felt that . . . I learned all of it quite well. The main ideas that we covered in the chapter were Quadratics, Polynomials, and Geometry-based problems.

The quadratics part of the chapter, I felt was the hardest. Because I missed the first day of that section, it was difficult for me to understand it. After I few days I understood what I was supposed to do. If it had gone any faster, then I think most students, including myself, would not have been able to follow what was going on.

The Polynomial part of this chapter was very easy in my opinion. It is just like normal equations with a little bit more challenge to it.

The geometry section of this chapter was a little more challenging than the polynomials. I liked this part of it because it involved working with shapes and figuring their areas, and I like doing those kinds of problems.

This chapter was overall a good one. I enjoy working out long problems, so I enjoyed this chapter. Some suggestions for years to come might be: a) keep the partner quiz and maybe do more, because it really helps you understand and helps others understand it; b) don't spend quite so much time on the simpler things and more time on the more complex things; c) do more group activities/projects to make it more interesting.

LeeAnne

Because journal writing is a communication of personal feelings and experiences, teachers may benefit from students' expressing ideas in their native language. Depending on your own knowledge, availability of translators and additional resources, this may be something you want to consider. Following is an excerpt from Rosaria's journal using the same prompt as Manuel and LeeAnne:

Este capítulo no fue tan dificil, solo que veces, no le entendía y veces no me sentía vien y por eso no ponía atención en lo que estaba asiendo pero ubo unos dias que se me asia todo muy fácil . . . Antes de que hiciéramos los polynomials, hicimos los quadratic y esos problemas se me hicieron un poquito porque no podía a ser la fórmula se me asia tan dificil que veses me tenía que llevar el cuaderno a mi casa para estudiarla pero por fin la aprendí.

Rosaria

Students enjoy journal writing. They view it as a way to express their feelings and even as a way to get personal in mathematics class. Unfortunately, some students see mathematics as a dry subject, one that requires them simply to memorize formulas and routines. With journal writing, students are given an opportunity to express their opinions about mathematical topics. Encourage students to write in their journals outside of class too, perhaps when they are doing their homework. Students will begin to realize that journal writing is a valuable opportunity to make mathematics more meaningful.

Chapter Four

Essays

What is an essay?

Probably the most common type of writing a student is expected to be able to do in high school is the formal essay. Students must be able to take a position on an issue and explain it or defend it in an organized and well-supported manner. Although journals are an excellent means of effective assessment, teachers also need to know that students are understanding the work in the mathematics classroom and can explain what they are learning. Essay writing is more effective for this type of evaluation. Many teachers include essay questions on tests to check students' understanding of ideas and their ability to explain and organize ideas quickly. Standard essays with an introductory thesis, body, and conclusion not only require math students to organize their ideas carefully, but also allow them to show what they know about the concepts they are learning.

The mathematics classroom is the perfect place for essay writing. Mathematics requires students to be logical thinkers—to analyze problems and use a step-by-step process to solve them. Essay writing is similar to problem solving in that it requires careful thought and logical organization to come to a reasonable conclusion. Students must know their purpose in writing, and they need to follow a writing format that is easily understood. The formal essay is the one writing style that fulfills all those needs because it can be used to explain, describe, persuade, or narrate. Essays are different from journal writing in that they require the writer to use coherent and unified writing to explain or prove an idea. Also, essays are evaluated both for the ideas expressed and for the way they are expressed.

An essay is not difficult to grade, provided the student has followed the teacher's guidelines for writing. Most students are familiar with the standard five-paragraph essay, although the teacher is certainly not limited to using only this style. The five-paragraph essay (introduction, three body paragraphs, and conclusion) allows for the teacher to check the introduction quickly for the thesis, look for the three main ideas the student is using to support his thesis in the body, and note the summary in the conclusion.

In the mathematics classroom, teachers need to discuss essay writing techniques if they want good results from their students. Before the first writing assignment is given, it is a good idea to explain the required essay format. Students need to know how they should organize their ideas, and many will ask "How long does it need to be?" Although length should never be the only criteria for an essay, the teacher may require the student to have at least three supporting ideas, or three body paragraphs.

Writing and grading essays

The teacher also needs to decide in advance how the essay will be graded. Some people may want to grade format and content equally, while others may weight one over the other. It is not recommended that you attach too much weight to minor editing issues such as spelling, punctuation, and capitalization. Focus on the ideas first—are they mathematically correct? Logical? Do they support the thesis? Next, examine the organization of the paper. Does the paper contain an introduction with a thesis? Are the supporting ideas arranged in a logical order? Does the conclusion restate the thesis and leave the reader with a final note of interest or thought about the issue? Finally, spelling, punctuation, and capitalization should be noted as the teacher deems necessary.

Depending on the assignment, teachers may want students to share their papers in small peer-editing groups before they turn them in. These groups would allow students to get useful feedback on their papers and allow them to make necessary corrections before they are graded. This also encourages students to correct careless proofreading errors. The members of a group may number anywhere from three to five. Students may have a checklist of the criteria for the assignment in front of them, and could mark the checklist for each paper they read. Students should then discuss the checklists and why items were marked as they were. For example, if a student received a no next to the item "Was the thesis statement clearly stated?" the student then has a chance to ask the others for help in revising the thesis. Peer-editing groups also allow students to share their understanding of the mathematics concept being discussed in the papers and may help to eliminate errors in thinking. Teachers may use essays as often as they like and have the time to grade. We believe in the importance of giving at least one essay question on every chapter test. Students need to practice essay writing frequently in order to become good writers and analytic thinkers. We also recommend assigning an essay at least once a month to allow students the opportunity to explore issues in mathematics.

Students have a difficult time writing good essays unless they have good questions with which to work. Harry Singer and Dan Donlan in *Reading and Learning from Text,* enumerate the three necessary components of a good essay question. "A well-written essay question contains three elements: (1) a *performance verb,* that is a verb that tells the student how to answer the question; (2) an indication of the *content* of the response; and (3) *enabling suggestions* for how to proceed in organizing the answer."[5]

Sample essay questions and student responses

The following are some sample essay questions and students' responses:

Question: In a well-organized essay, write about the importance of the point-to-point pattern in graphing. Your essay should include an explanation of the following topics:

a) What happens if the point-to-point pattern remains the same throughout the entire graph?

b) What kind of graph is created by changing the pattern as you move from point to point?

c) What does the graph of a line look like if it has a positive slope? a negative slope? a slope of zero? an undefined slope?

d) What does the graph of a line look like if its slope is far from zero, for example, 100 or –100? What if its slope is close to zero, for example, 0.01 or –0.01?

e) What is special about the slope of parallel lines?

The point to point pattern is very important in graphing equations. It is, what one might call, the "Link" to finding other solutions to an equation. If the point to point stays the same throughout the graph, then you know you have done the problem correctly you now have a series of answers to put in your equation. By changing the pattern as you move, you create a curving line rather than a straight line and your points are spread out a little more. If your graph has a positive slope it will slant up to the right but if it has a negative slope it will slant down to the right. If your slant is zero, you will end up with a horizontal line, but if its undefined, you will end up with a vertical line. If a line's slope is far from zero, it will be nearly a vertical line but if it's close to zero, it will be nearly a horizontal line. Parallel lines are very interesting. They are two lines that go side by side and they never touch, but they also have the exact same slope, or point to point pattern.

Linda

Question: Write a well organized essay that discusses the purpose of solving a system of equations in two variables. Include a description of the methods for solving a system. Somewhere in the essay, answer the following questions:

a) What would the graph of a system look like if the system had no solution? one solution? infinitely many solutions?

b) What is the role of the slopes in helping you determine the number of solutions to the system?

c) What is the role of the y-intercepts in helping you determine the number of solutions to the system?

d) If you are using the substitution method, or linear combinations method, how can you tell if there is no solution, one solution, or infinitely many solutions?

e) Explain some of the benefits and drawbacks to using a technology approach.

The purpose of solving a system of equations is to find the number or numbers that would work in place of the variable. You could do it many ways; by the addition method, where you line up the equations to add them together to get the first variable then do it like the substitution method. In the substitution method you solve for "y" then substitute that into the other equations to get "x." The graphing method is where you solve both equations for "y" and graph each line. The linear combinations method is where you get opposites after solving for "y" and line up the equations, then do it like to addition method. On the graphing calculator you graph each line and "trace" to the intersection.

If the system had one solution the two lines would cross at one point once. If there were no solutions the lines would be parallel. If there were infinitely many solutions they would be the same line. The slopes help you determine the number of answers by telling you if the lines will meet once, never meet, or be the same line. The y-intercept tells you where the line crosses the y-axis, and where to start your slopes, and if they'll cross.

If you're using the substitution method or L.C.M. you can tell how many solutions there will be by solving for "y" and seeing if the slope, and y-intercept are same or different. Some of the benefits of using the calculator is that it can be done faster, is easier to see the answer, and can do decimals easier. Some of the drawbacks are that it can be complex, it's not always perfect, you have to solve for both "y's," and you don't always have a calculator with you.

Joseph

In Algebra we sovle systems to find were two lines cross and which points will work in both eqution When sovling a system you should sovle for y= first but if it is adition you will only have to make sure you have compatble numbers if not times it by something to get the numbers. We have many differnt looking graph's. Ocasional we have a graph with one soltion in graph like that it would look like two lines crossing each other at one point in time. But some times we have no soltions this would look like to line running with each other but never tocching. They are parala and we have a graph with many which look like two lines running on top of each other. One way of helping us determine this before we graph are the slopes or point to point pattern. If they are diffrent we will have many soltions. The y-intercept will tell us if the is just one point or if no points. It will tell us if they are on top or not. But if you use linear or substitution you look the x and y equal to see the y-intercept and the number with the x for the point to point pattern you also can tell if ther are many soltions by if the point to point the same of diffrent.

There are some drawback to technology tho. but also benefits as long as you solve for y= then you punch it in your calculater when tryng this you hit graph y= then put it in the problem, hit the : then graph y= again after problem is in ther EXE. The problem will come on the screen and by modern technology you can find the x= and the y= by tracing the intercept ot the lines the drawbacks are if you don't have one or no how to opprate a calculate you can't use it and not always is it ther and you could have trouble finding the crossing point of the problem benefits are it's quick, painless no witing and usally very easy to read, so in the fast pace of Bussnes would solving eqution importan and hopefully know you can tell if your graph has no many or one solution

Jonathon

Each of these student pieces is priceless in that it provides great insight into the student's understanding. These essays illustrate the type of information that can be obtained from students using writing instead of the traditional problem-answer format.

In advanced classes, the essay is an invaluable tool to help students organize, structure, and personalize new knowledge. Following are some excerpts from a six-page essay submitted in a pre calculus class in response to the following direction:

Question: In a well organized essay, describe the procedures one would follow in order to graph *any rational function* without the aid of a graphing utility. Explain the concepts as well as the mechanics of any situation that you describe. Be sure to include advice on all areas of potential difficulty. Use examples when needed to illustrate ideas or techniques.

A rational function is a function that has a polynomial in its numerator and denominator.

$$\frac{(x+a)}{(x-b)}$$

is an example of a rational function. Rational functions are unique because of their complexity. They often include both horizontal and vertical discontinuities and asymptotes. Graphing a rational function without a graphing utility can be rather tricky, but following some simple steps can make the task seem rather simple . . .

Langston

To start off with it is helpful to know what occurs at the end behavior of the graph. With a rational function such as this it is accurate to take the highest powered numerator [term] and divide it by the highest powered denominator term. In my example, I end up with

$$\frac{x^2}{x^4}$$

which simplifies to

$$\frac{1}{x^2}$$

This means that as the X values get incredibly large in absolute value the graph will behave like

$$\frac{1}{x^2}$$

Basically graphs will have one of three types of end behaviors. If, when simplified, the function has a X term left in the denominator (like my example did) then as X approaches infinity f(x) will approach 0. If, when simplified, the function has no X terms left there will be a fraction left by the coefficients of the X terms. This fraction is what f(x) will approach as X goes to infinity. The last type is when there remains a X term in the numerator after simplification. When that happens f(x) will approach infinity as X approaches infinity.

Daniel

As this chapter ends, we offer some additional essay questions that have been used successfully in the classroom:

- Throughout the chapter on radicals, some students continue to make the same mistakes over and over again. It seems as though they are not studying. On the other hand, they could be getting things confused in their minds. Below is an old quiz with somebody's answers written on it. Next to each problem either indicate the answer is correct or write some comments as if you were the teacher trying to correct the misunderstanding.

  a) $\sqrt{9c^6} = 4.5c^3$

  b) $\sqrt{121x^{20}y^{16}z^4} = 11x^{10}y^8z^2$

  c) $\sqrt{256} = \sqrt{16} = \sqrt{4} = \sqrt{2} = 1.4142135...$

  d) $\sqrt{144x^2} = 72x^{1.4142135...}$

  e) $\sqrt{16x^{36}} = 4x^6$

- As you prepare for your test on radicals, you should be able to understand the difference between some operations on radicals. For each of the following questions, explain in words the difference between the two concepts. Then illustrate each operation with an example.

  a) What's the difference between simplifying a radical and giving a decimal approximation?

  b) What's the difference between multiplying radicals and adding radicals?

- Describe what we mean in algebra by term. What are like terms? When can terms be combined? How are terms combined? Why do we combined terms in an algebraic expression whenever possible? How are terms multiplied? What is the difference between adding terms and multiplying terms? Be sure to use correct terminology and illustrate your points with examples.

- Given below are three different methods of recording a set of the same test scores. Compare and contrast the three methods and the different information you can get from each.

  method 1: 55, 68, 85, 89, 58, 95, 53, 86, 100, 85, 88, 65, 88, 89, 41, 66, 98, 95, 99, 93, 0, 47, 87, 0, 66, 37, 46, 85, 79, 89

  method 2:

0	0 0
1	
2	
3	7
4	1 6 7
5	3 5 8
6	5 6 6 8

Stem	Leaves
7	9
8	5 5 5 6 7 8 8 9 9 9
9	3 5 5 8 9
10	0

method 3:

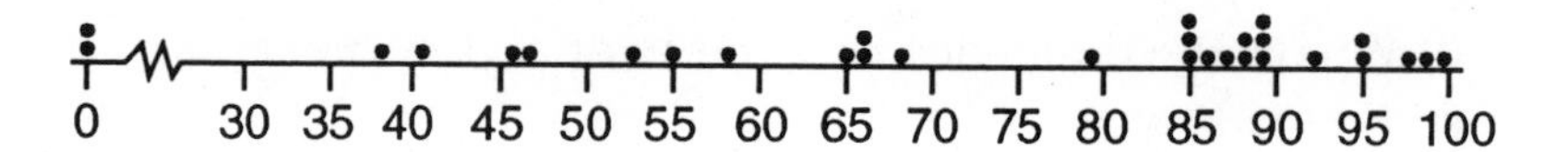

☛ In class we recently learned about the Pythagorean Theorem. Take this opportunity to create your own Pythagorean Theorem word problem. For full credit, your work must contain these four parts:

a) A description of the problem without a picture. All the information necessary to solve the problem should be included in the word description.

b) A diagram that you would use to solve the problem.

c) An explanation of why you believe you can solve it using the Pythagorean Theorem.

d) Your solution to the problem, with all work shown.

☛ One night last week Jim was working on his factoring homework at the kitchen table. The phone rang and he left to answer it. When he returned, he found his younger brother had spilled finger paint all over his homework. Below is a copy of Jim's paper. Determine, based on what you can see, the possible solutions to the assignment. Clearly explain your reasoning.

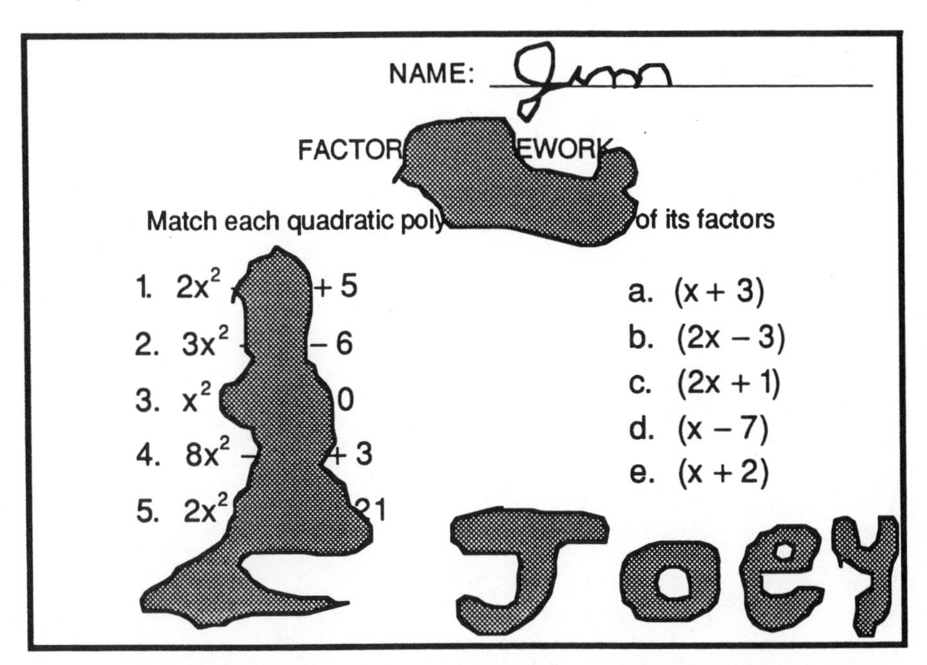
NAME: Jim

FACTOR[illegible]EWORK

Match each quadratic pol[illegible]of its factors

1. $2x^2$ [illegible] $+ 5$
2. $3x^2$ [illegible] $- 6$
3. x^2 [illegible] 0
4. $8x^2$ [illegible] $+ 3$
5. $2x^2$ [illegible] 1

a. $(x + 3)$
b. $(2x - 3)$
c. $(2x + 1)$
d. $(x - 7)$
e. $(x + 2)$

Joey

- Summarize the behavior of real-valued linear functions under iteration. Within your summary include examples and explanations of the various behaviors that can be observed. Use appropriate vocabulary.

- In a well-organized essay explain why many people, including some mathematicians, find the work of Mandelbrot and the ideas surrounding fractal geometry unacceptable. Include in your essay, but do not limit yourself to, the following questions.

  a) What developments within the mathematical sciences occurred at the end of the 19th century and continued into the 20th century?

  b) With the advent of computing tools, how did the study of mathematics change?

  c) Is there a need for mathematical rigor and proof? Justify your position.

  d) What are some of the ways in which fractal geometry has been useful?

Chapter Five

Research and Project Writing

One of the more exciting writing assignments we can give students is research writing, where students gather and analyze ideas from sources outside the classroom. What makes research exciting is that students make connections between what they learn in class and what they learn from other sources. When students ask questions such as, "What does this have to do with the real world?" or "When will I ever use this in the real world?" they are revealing a lack of awareness of these connections.

Through research and project writing, students develop skills that are important to their performance in the work place. Carnevale, Gainer, and Meltzer in their publication *Workplace Basics: The Essential Skills Employers Want*[6] state that the following seven skills are to be considered basics for the 1990s:

1) Organizational Effectiveness and Leadership
2) Interpersonal Skills, Negotiation, and Teamwork
3) Self-esteem, Goal Setting, Motivation, Personal and Career Development
4) Creative Thinking and Problem Solving
5) Communication—Listening and Oral Communication
6) Reading, Writing, and Computation
7) Learning to Learn

In the working world people rarely work in isolation. Nor do people work on projects unrelated to anything else. For example, as teachers we often work together with other teachers to solve a problem, write a curriculum, or develop a lesson. Many teachers also serve on committees, in organizations, and on commissions where they must work with others to write plans, reports, and letters.

Research and project writing allow students to work on issues related to the real world and to work with others in cooperative groups when appropriate. Heterogeneous grouping is often effective in giving everyone a chance to learn. Give each student a specific task, such as recorder or reporter, and allow students to work together to solve their problem. Schedule class time for students to work in their groups or to gather information in the library as needed. Librarians can be very helpful in gathering specific reference materials for students working on a particular project.

Some of the more successful projects students have done have been those that allowed them to explore applications of mathematics outside the classroom. In algebra II students were asked to interview a person in a profession they might be interested in pursuing as a career themselves. The students then wrote papers and reported to the class on the amount of mathematics needed to get that job as well as the mathematics used on the job. In algebra I, students interviewed family members about the mathematics they used in everyday life or on the job. In conducting these interviews, many students began to realize what may be expected of them in the professions of their choice. They should be encouraged to explore a wide array of occupations, including doctors, dentists, engineers, assembly line supervisors, computer operators, secretaries, architects, and nurses. Students and teachers alike are often surprised to learn how great a role mathematics plays in all our lives.

In a geometry class students can be asked to research and report on a variety of applications to daily life. Examples of similarity, right triangle trigonometry, transformational geometry, tessellations, congruence, and proof can all be found in your local community. A particularly interesting project for geometry students requires them to develop their own postulate systems for a given situation. From these postulates, one theorem must be deduced that can be proved. Below is one student's submission.

The Movie Postulates:

1) If I watch the previews then I will get bored.
2) I will have money iff I go see a movie.
3) If I eat popcorn then I will get thirsty.
4) If I buy a soda then I will drink it.
5) If I get sleepy then I will fall asleep.
6) If I have a ticket then I will be admitted to the movie.
7) I do not have money iff I do not buy a ticket.
8) If I watch the previews then I will get hungry.
9) If I snore then I will get kicked out of the movie.
10) If I fall asleep then I will snore.
11) If I do not have to go to the bathroom then I did not drink a soda.
12) If I am admitted into the movie then I will have to watch the previews.
13) If I get thirsty then I will buy a soda.
14) If I get popcorn then I will eat popcorn.
15) If I get bored then I will get sleepy.
16) I do not get hungry iff I do not get popcorn.

Theorem: If I go see a movie then I will have to go to the bathroom.

Leo

From the structure of these postulates, it becomes apparent that the student is familiar with conditionals, bi-conditionals, inverses, contrapositives, and converses. Furthermore, the student has illustrated a depth of understanding that would be very difficult to assess with an objective test.

Another project that can be used at the algebra I level requires students to use what they know about interpreting graphs. Students are put into groups of three or four, and each group is assigned a non linear graph. The graph is presented to them on a set of axes without labels or marked intervals. Each group is given the task of creating a situation that could be represented by the graph. As part of a typical discussion, students will use mathematical terms such as positive slope, x-axis, and y-intercept to assign meaning to the characteristics of the graph. Students are then asked to write a news story that might accompany their graph in a newspaper. This gives each group an opportunity to communicate the details of their created situation to others. Lastly each group has the task of presenting to the rest of the class a brief presentation that represents the origins of their graph. Assigned this task, students have responded by acting out board meetings at famous corporations and dramatizing the plight of a celebrity on a diet.

Another group project that has brought wonderful results is the mathematical analysis of excerpts of Jonathan Swift's *Gulliver's Travels.* To begin this project, read the following excerpt from the novel.

"Two hundred Sempstresses were employed to make me Shirts. . . . The Sempstresses took my Measure as I lay on the Ground, one standing at my Neck and another at my Mid-Leg, with a strong Cord extended, that each held by the End, while the third measured the Length of the Cord with a Rule of an Inch long. Then they measured my right Thumb, and desired no more; for by a mathematical Computation, that twice round the Thumb is once round the Wrist, and so on to the Neck and the Waist; and by the Help of my old Shirt, which I displayed on the Ground before them for a Pattern, they fitted me exactly. . . .

I had three hundred Cooks to dress my Victuals . . . , and prepared me two Dishes a-piece. . . . A Dish of their Meat was a good Mouthful, and a Barrel of their Liquor a reasonable Draught."[7]

From this, students are to determine the accuracy of Swift's statements, assuming that, as he had previously said, the Lilliputians were less than six inches high. Each group then has the task of rewriting the excerpt and correcting any miscalculation.

Precalculus students, too, love to do projects, but at times we forget to spice up their learning experiences. Divide students into groups of two or three and pose the following problem.

Project: Imagine that you have been hired by an oil company to assist them in keeping their costs low. Given the following diagram, find the single location on the pipeline to which you should connect all three oil wells so as to minimize the cost of hooking up. Then write a letter to the president of the company, explaining in detail your recommendation.

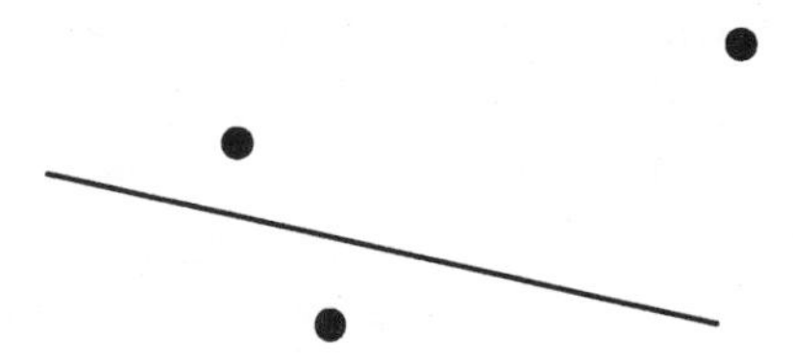

Writing projects can also be based on media presentations. Ask students to watch a NOVA documentary and write an essay afterwards detailing their reactions and newly acquired knowledge. Such programs are particularly effective in demonstrating the applications of mathematics. Writing gives students an opportunity to assimilate and reflect on the new information. Below are one student's remarks after viewing a program.

I don't know about everyone else in this class, but fractals and chaos really intrigue me. They link up with so many things in nature. I would never thought such things that seem so unpredictable, could have an air of predictability to them. . . .

However Newton had a totally different view of the universe. . . . He felt that the future could be predicted if only enough information about the present situation was known. . . . Edward Lorenz can be credited with one of the breakthroughs in chaos theory, 200 years after the acceptance of Newton's beliefs. You never know what scientists will discover in the next 200 years. . . .

All of these things that we have done in chaos and fractals have really caught my interest. I find it absolutely fascinating that such things can be connected. It makes me wonder what else is out there, ready to be discovered and studied by someone like me.

Manisha

The NCTM *Standards* call for making more connections between mathematics and the modeling of real-world phenomena. Research and project writing afford students the opportunity to examine these applications. Recording the discoveries of new ideas and their reactions to them is often a rewarding experience for students. Equally enjoyable is the role of teacher as observer, model, and coach in the development of very important skills in our students' lives.

Chapter Six

Evaluation of Writing

There are numerous approaches to the evaluation of student writing. In this chapter we present a variety of methods and encourage teachers to use those that are best suited to their needs.

Portfolios

One approach to evaluation is the creation of a writing portfolio. At the beginning of the year, students are given a folder in which they keep all writing they do in the class. Periodically, the teacher asks the student to examine the work in the portfolio for the purpose of selecting the five best pieces of work for the year. This number may vary, depending on teacher preferences. Students may be given the option to further revise this work or may turn it in as is. Teachers may have already graded the work or may have simply marked it complete previously and will now evaluate it.

The portfolio provides a broad perspective of the kinds of writing the student has been doing in the classroom. Giving students an opportunity to evaluate and revise their work before turning it in for a grade encourages critical thinking about their writing and about what they have accomplished in the classroom.

If all teachers in a given subject area choose to use portfolios, then it becomes a valuable resource for the teacher a student has the following year. A mathematics portfolio that has been maintained and used during the four years of high school gives the student dramatic evidence of growth in mathematical power.

Another portfolio option allows the student complete freedom to decide what becomes part of the portfolio. Students should be advised early in the year that they will be expected to collect several pieces of their work in mathematics. Encourage students to choose a variety of samples that will provide the reviewer with a complete picture of a student's mathematical growth and knowledge. Parents may view the portfolio during conference night and may also be asked to write their own observations of the student's progress for inclusion. This will give some insight into how much mathematical growth is perceived at home. Following are some suggested pieces of work that could be included in a portfolio:

1) Five consecutive days of notes

2) Five personal journal responses to different lessons in a quarter

3) Five consecutive homework assignments with reactions to the problems

4) Two examples of a student's work from a group project

5) Two responses to NOVA videos

6) Two responses to a magazine or newspaper article where mathematics was mentioned and used

7) Evaluations that fellow group members have written about a student at the end of a group project

8) An original solution to a problem that was worked on in class

Peer Evaluation

Another type of evaluation that directly involves the students is peer evaluation. Students work in writing groups or with a partner, and together they read a draft of each composition and revise it based on criteria suggested by the teacher. It is most helpful for the teacher to provide the students with a list of requirements, including items such as a thesis statement, supporting ideas, transitions, and proofreading marks.

The peer evaluators are able to give a writer instant feedback and ask for clarification of ideas as needed. Each writer is able to make corrections to the assignment and turn it in later, at which time the teacher may decide how to grade it. We do not recommend allowing students to grade each other's papers. The peer evaluation group is meant to offer a chance for students to share their writing, get feedback on their ideas, and revise as needed to create their best work.

Using Checklists

Evaluation can become more focused through the use of a checklist. With this method, the teacher considers the requirements of the writing assignment and uses a checklist while reading the paper to help determine whether the requirements have been met. The teacher may focus on a few criteria or create an extensive list of items for students to include in their papers. Copies of the checklists may be given to the students as they receive the assignment, or when they are evaluating their own and each other's papers. As the examples below show, a variety of checklists can be created to meet different objectives. We have included several examples, some of which are used in English classes at West Chicago Community High School.[8]

The first checklist provides a specific list of the criteria by which the teacher evaluates the student papers. Many students will find it helpful to have a copy of this while they are writing their assignment, and especially while they are revising their papers. Furthermore, the checklist helps the teacher to be consistent when grading papers and provides a quick reference for the teacher when assigning grades.

Checklist 2 is similar to the first one in that it also provides a listing of the specific requirements of a particular assignment. The design of this checklist allows the teacher to make comments in the right margin detailing strengths or weaknesses in the student's paper. Of course, for any checklist students need to have each of the criteria carefully explained to them before they can begin to use them to write and evaluate papers.

The third checklist is written in a more traditional form. Possible scores (1–5) are listed at the top, with individual scores to be written at the far right of each category. This particular format allows certain categories (reasoning, organization, content, and supporting paragraphs) to carry more weight than others, and could be adapted to suit a particular assignment. With this format the student receives feedback on both writing skills and understanding of the material, and the teacher saves some time in communicating that feedback on the student's paper.

For group projects and class presentations, Checklist 4 has been used successfully. It allows for a blend of individual assessment (for example, Ability to follow directions) and group assessment (for example, Quality of final output).

The last checklist is a scoring chart that can be used to write specific descriptions for each possible point value on an assignment. This particular one accommodates six criteria with four possible point values in each. The chart may be expanded or contracted to fit a specific application.

Checklist 1

Total Score: ____

	Excellent (4)	Good (3)	Fair (2)	Poor (1)
1. Specific, arguable thesis	____	____	____	____
2. Support of thesis, use of logical arguments	____	____	____	____
3. Concrete facts & examples	____	____	____	____
4. Logical progression of ideas	____	____	____	____
5. Coherence or flow of ideas	____	____	____	____
6. Unity	____	____	____	____
7. Clarity/clearness	____	____	____	____
8. Paragraph development	____	____	____	____
9. Effective sentence structure & variety	____	____	____	____
10. Diction & use of language	____	____	____	____
11. Grammar & correctness	____	____	____	____
12. Worthiness of subject	____	____	____	____
13. Persuasiveness or ability to convince reader	____	____	____	____
14. Originality & overall interest	____	____	____	____

Checklist 2

Total Score: ______

	Yes	No	Comments
1. Did the writer compose an introductory paragraph that includes a thesis statement?	____	____	____________
2. Did the writer state the topic and purpose clearly in the thesis?	____	____	____________
3. Did the writer develop one aspect of the topic in each body paragraph?	____	____	____________
4. Did the writer limit the details within each paragraph to those that develop the main idea of that paragraph?	____	____	____________
5. Did the writer clearly organize his or her ideas?	____	____	____________
6. Did the writer make clear transitions between ideas?	____	____	____________
7. Did the writer bring the essay to a definite close in a concluding paragraph?	____	____	____________
8. Did the writer choose a suitable title for the essay?	____	____	____________
9. Did the writer use correct grammar, usage, spelling, punctuation, and capitalization?	____	____	____________
10. Did the writer carefully proofread the final copy?	____	____	____________

Checklist 3

	1	2	3	4	5
Introduction	Very poor. You haven't interested the reader.		Some techniques used to try to catch the reader's attention		Very well done! You grabbed the reader! — score times 1 = ______
Thesis Statement	Not well stated for a valid argument.		Has good possibilities but still needs to be improved.		Well stated, aptly explains your valid argument. — score times 1 = ______
Reasoning	Not logical or was unable to be followed.		On right track but sometimes confusing.		Very well applied in this argument. — score times 4 = ______
Organization	Not in logical order or not including all parts.		Some parts well done; others poorly handled.		Very well organized All parts included. — score times 4 = ______

Checklist 3 (cont.)

	1	2	3	4	5
Supporting Paragraphs	Poorly written. Sentence didn't flow, weren't organized, or didn't use transitions.		Some paragraphs were well written, but others needed help.		Good paragraphs with strong sentences and useful transitions. score times 4 = ______
Content	Several mathematical errors. More study recommended followed by revision.		A few mathematical errors. Re-read your notes and revise them.		Excellent content. Explanations and connections very well made. score times 5 = ______
Conclusion	Doesn't sum up main reasons presented in support of proposal or show how those reasons led to your proposal.		A good attempt, but needs work.		Sums up main reasons in support of proposal or shows how those led to proposal. score times 1 = ______

Total score = ______

Checklist 4

Total Score: ______

Part One: Preparation (50 points)

1. Cooperation in groups as observed by the teacher (15 points maximum) ______
2. Ability to follow directions (10 points maximum) ______
3. Quality of individual work / input to group (15 points maximum) ______
4. Quality of group output (10 points maximum) ______

Part One Total: ______

Part Two: Presentation (50 points)

1. Effective use of preparation time (15 points maximum) ______
2. Relatedness between graph and story (15 points maximum) ______
3. Creativity / Originality (10 points maximum) ______
4. Work divided equally among group members (10 points maximum) ______

Part Two Total: ______

Checklist 5

Scoring Scale			
Descriptors: Upper Half		Descriptors: Lower Half	
4	**3**	**2**	**1**

Holistic Evaluation

A teacher may evaluate a paper holistically, or as a whole, rather than by marking for each of several criteria. There are different ways to do this. One way is simply to grade one paper at a time, evaluating the paper as whole. The teacher reads through the entire paper, making comments as needed, and then assigns a grade. If the overall result adequately meets the teacher's requirements, the student would receive a C or better, depending on the quality.

Following another method, the teacher reads through all the papers before assigning any grades, writing comments on each and putting the papers in order, from best to worst. Having done this, the teacher then goes back to each paper and assigns a grades. This method allows the teacher to determine whether the assignment was clearly stated, and to make sure that all students completed it correctly. Students who mastered all requirements of the writing assignment would receive an A, and other students would be marked accordingly. Teachers who prefer grading on a curve might find this evaluation method helpful.

Holistic grading allows the teacher to move through papers quickly, making it a convenient evaluation tool for essay test questions. However, the teacher still needs to know the precise criteria for grading. Furthermore, when a grade is given, the comments should indicate clearly why the student received that grade. Students appreciate having both their strengths and weaknesses pointed out to them. In addition, students might be allowed to revise their papers for a higher grade after having had a chance to read through the comments and perhaps discuss the paper with the teacher. Learning to write well is a never-ending process, and students benefit from having the opportunity to work on their writing even after it has been graded.

No matter what approach you choose to use in evaluating the writing done in the mathematics classroom, keep in mind that the overall focus should always remain on what the student is gaining from the writing exercise. Portfolios, peer evaluation, checklists, and holistic grading must all keep this focus. The evaluation process must not put an end to creativity and productivity but rather serve as an incentive to help the student improve. Effective evaluation of writing will foster growth in mathematical thinking.

Chapter Seven

A Sample Algebra I Unit

This unit was designed to make the topics more interdependent and coherent in the eyes of the students. The writing tasks add greatly to the students' understanding and retention of the material. During this unit students are exposed to direct instruction, cooperative learning, discovery learning, independent research, and three types of assessment strategies—journals (Day 1), exhibition (Days 13–15), and testing (Day 16).

Outline for Unit on Graph Analysis

Unit Objectives: As a result of this unit, students should be able to

a) Graph a linear equation in two variables.

b) Use a linear modeling equation to predict outcomes.

c) Analyze predicted outcomes for reasonableness and, if possible, correctness.

d) Construct a scatterplot to compare two sets of data.

e) Analyze a scatterplot using the line of best fit.

f) Use their understanding of linear graphs to analyze non-linear graphs.

g) Draw an approximate graph of a situation given a verbal or written description.

Day 1: Introduction to the Cartesian Plane

This first day uses a general review to reacquaint students with the Cartesian coordinate system. Students are asked to write a reflective journal entry, which gives the teacher immediate feedback so that the next lesson can be formatted in an appropriate manner. The journal is a response to the following questions:

- What were the goals in today's algebra class?
- How does this connect with what I knew before?
- What did the teacher do today that helped me learn?
- How do I feel about today's algebra class?

Day 2: The Cartesian Plane

Students work in pairs to familiarize themselves with the plane. Each student must describe a picture superimposed on a plane, using only numbers and the words *up, over, down, left,* and *right.* The partner attempts to recreate the picture based on the description.

Journal entry: What are your reactions to today's activity? How does it relate to analyzing graphs?

Day 3: Introduction to Linear Equations in Two Variables

Each student is assigned an ordered pair of numbers indicating their position on the coordinate plane representing the classroom. All students whose ordered pairs satisfy a given linear equation stand up. The teacher then asks questions such as, "What is the desk-to-desk pattern?" or "Where would the next person have to sit in order to satisfy the equation?" The desk arrangement as it relates to the Cartesian plane becomes a way to identify all possible desks. Students then realize that after identifying two points as solutions, a point-to-point pattern—the slope—can be defined.

Students are asked to write a paragraph for homework in which they describe the relationship between the desk-to-desk pattern and the cursor pattern on the calculator, identifying possible explanations for the connection.

Day 4: Graphing Lines Using Tables

This lesson is presented in a direct instruction format stressing the point-to-point pattern as a means of verifying the points generated in the table.

Journal topic: How can patterns be used to predict things in your life?

Day 5: Introduction to the Slope-Intercept Form

The format of this lesson is guided discovery. The lesson ties together the ideas of point-to-point pattern and plotting points. Students gather data by graphing three linear equations. After the graphs are generated, students work in groups to hypothesize the connections between the points they have plotted and the terms of the equation. By the end of the lesson each student should know how to graph a linear equation in $y = mx + b$ form.

Day 6: Using the Slope-Intercept Form

Students learn the terminology and practice graphing equations using the slope intercept form.

Day 7: Using the Slope-Intercept Form

Students are asked to write an equation of a line given two points. Some use the idea of point-to-point pattern to locate the y-intercept, while others find very creative ways to generate the correct equation. Each group must provide a written explanation of their proposed method with examples worked.

Day 8: Using Equations to Model Situations

Students work with a variety of real-world situations where specific outcomes can be determined using a linear equation. For example:

- The cost of attending a movie can be found using:

  $$C(x) = 6.75x$$

  Where x is the number of people, over six years of age, attending the movie.

- The price of a plumber's house visit can be predicted using:

  $$\text{Price} = \$42.50\ (\text{number of hours}) + \$25.00$$

  Where $25.00 is a charge just for coming and he then charges $42.50 per hour of labor.

Students see that these equations are algebraic models of the described situation. They are asked to write a paragraph on how predicting outcomes using linear models uses their equation solving skills.

Day 9: Using Equations to Model Situations

Students work in groups to think of situations that can be described as a linear relationship. If needed, they are encouraged to use the library to research their situation.

Day 10: Scatterplots and Best-Fit Lines

Students explore the scatterplot as a tool to investigate possible relationships. They collect data on the circumference and diameter of circular objects and graph it on a scatterplot. After being introduced to the concept of line of best fit, students are asked to write a paragraph on the following: "Why is the line of best fit an acceptable approximation?"

Day 11: Scatterplots and Best-Fit Lines

Students work in pairs to generate data for a scatterplot by measuring their body height and navel height. The class then determines a line of best fit and generates a modeling equation for the data. Have students use the equation for making predictions and discuss their reasonableness.

Research topic: Research why the Greeks called the slope of this line the Golden Ratio.

Day 12: Introduction to Nonlinear Graphs

The students use their calculators to generate tables of points for a variety of functions that are unfamiliar to them, such as $y = \sin x$, $y = \tan x$, $y = \sqrt{x}$, $y = x^2$. They then graph the points and discuss the behavior of non-linear graphs.

Days 13–15: Assessment: Interpreting Nonlinear Graphs through Storytelling

This project is designed as the culminating activity for the graph analysis unit. Students are assigned to work in groups of four. Each group is assigned an unlabeled graph to analyze. Unlabeled quadratics, absolute value curves, cubics, quartics, logarithm curves, or any other curves may be used. The group members must agree on a real-world situation that could be modeled by their assigned graph. On one copy of the graph, the group imposes a coordinate system, calibrates the axes appropriately, and labels all important features. The group then writes a short story describing the situation that generated the graph. Extra points are awarded for including statistics that they have researched. After writing the story, the students develop the script for an in-class performance of their short story. The graph is to remain the focus of the skit. During the videotaped performance students are required to include a presentation of the graph and an explanation of its importance. Students are evaluated using a pre-written checklist. They are given a copy of the checklist beforehand, so they know exactly what is expected of them.

Day 16: Assessment

Students are given a traditional test to assess skills. Included on this test is the following essay question:

In a well-organized essay, write about the importance of the point-to-point pattern in graphing. Your essay should include an explanation of the following topics:

a) What happens if the point-to-point pattern remains the same throughout the entire graph?

b) What kind of graph is created by changing the pattern as you move from point to point?

c) What does the graph of a line look like if it has a positive slope? a negative slope? a slope of zero? an undefined slope?

d) What does the graph of a line look like if its slope is a number far from zero, for example 100 or –100? What if it is close to zero, for example 0.01 or –0.01?

e) What is special about the slope of parallel lines?

Chapter Eight

Writing Topics for Middle School

Examples thus far have been focused on topics for writing at the high school level, but there is also a need for writing in the middle school mathematics classroom. Therefore this chapter offers some suggestions for journal, essay, and research topics that are appropriate for grades six through nine.

Whole Numbers

- Find the altitude of your city and do some research to find out how it is calculated.
- Which is easier for you—multiplication or division? Why do you think one is easier than the other?

Decimals

- Explain the difference between truncation and rounding. Which one do you prefer? Give examples of each and situations where each is appropriate.
- Describe situations in your life when you need to use decimals.
- Fred is very upset. He multiplies 5 by 3, and gets a greater

number—15. But when he multiplies 0.5 by 0.3, he gets a number that is less than either one—0.15. He has always thought that multiplying two numbers results in a number greater than either of the two factors. Explain to him why the results can be different when he multiplies decimals. In your explanation, include examples of when the product of the two decimals lies between the two factors and when the product is greater than each of the two factors.

Metric Measurement

- Write a letter to the President of the United States, convincing him that the United States should or should not convert to the metric system.
- Research the origins of the metric system.
- Write a note to your mother, explaining why it would be better for her to spend $1.99 on a two-liter bottle of soda than on a six-pack of soda.

Equations and Number Theory

- Explain the need for the order of operations. Give an example of how a calculation could have several different answers if the order of operations is not followed.
- Do you think there are a limited number of prime numbers or an infinite number? Explain your reasoning.

Fractions

- Explain in three different ways what it means when fractions are equivalent.
- Why do you need a common denominator to add fractions?

Introduction to Geometry

- Susie is the editor of the yearbook. She is working with a picture that covers 100 squares on her grid paper. She notices that after she cuts each side in half, the picture only covers 25 squares. This is a problem for her because she needs to cover 50 squares with this picture. She doesn't understand why this is happening, and she is becoming very frustrated. Write a note to her explaining why this is happening and give her some hints on how to solve her problem.
- Your parents have asked you to help redecorate your house. They want to put new carpeting in the living room, new wallpaper in the kitchen, and an in-ground pool in the yard. Determine the measurements they need and prepare a report for them on the anticipated cost of this project.
- You have been hired by the Johnsons to design a garden that contains 48 square feet of planting area. Give them examples of at least four different rectangular gardens. Include in each the amount of fencing required to surround the garden and how you calculated the area.
- You have a piece of wrapping paper that measures 2 feet by $1\frac{1}{2}$ feet. Give the dimensions of three different boxes that could each be covered with this piece of paper with a minimum amount of waste.

Ratio/Proportion/Percent

- If two boys drink six bottles of soda in two days, how can you determine the number of bottle ten boys will drink in eight days?
- Clearly explain how a city with 9.2 persons per square mile could have a smaller population than one with 5.6 persons per square mile.
- You open a music store, and you decide to reduce all of the prices in your store by 10 percent in order to attract customers. As a result, your sales increase. Now you decide to

raise all the prices by 10 percent. Discuss whether raising the prices is a smart business move and why.

Statistics and Probability

- Keep track of the progress of your favorite baseball team over the course of a month. Use that information to predict its performance for the rest of the year. Justify your prediction.
- You have a weekend curfew of 10:00 PM. For twelve weekends in a row, you have come home by 10:00. However, on the Friday of the thirteenth week, you arrive home fifteen minutes late, and your parents ground you for the next two weekends because they think you will be late again. Write them a note in which you use your knowledge of statistics to convince them that it is not likely you will be late again.

Integers

- Explain the multiplication rules for positive and negative numbers.
- When might you use negative numbers in your life?

Notes

[1]National Council of Teachers of Mathematics, Commission on Standards for School Mathematics, *Curriculum and Evaluation Standards for School Mathematics* (Reston, Va.: The Council, 1989), p. 233.

[2]Sam Crowell, "A New Way of Thinking: The Challenge of the Future," *Educational Leadership,* 47 (1), p. 60.

[3]Jean Kerr Stenmark, *Assessment Alternatives in Mathematics* (Berkeley, Calif.: California Mathematics Council Campaign for Mathematics and EQUALS, 1989), p. 2.

[4]Judith L. Schwartz, "The Intellectual Costs of Secrecy in Mathematics Assessment," *Expanding Student Assessment,* ed. Vito Perrone (Alexandra, Va.: Association for Supervision and Curriculum Development, 1991), p. 137.

[5]Harry Singer and Dan Donlan, *Reading and Learning from Text* (Hillsdaler, N.J.: Lawrence Erlbaum Associates, 1980), p. 148.

[6]Anthony P. Carnevale, Leila J. Gainer, and Ann S. Meltzer, *Workplace Basics: The Essential Skills Employers Want* (San Francisco: Jossey-Bass Publishers, 1990), p. 3.

[7]Jonathon Swift, *Gulliver's Travels* (New York: Signet, New American Library, 1960), pp. 74–76.

[8]Kim Austin, Dorothy Carter, Tim Courtney, Wayne Kosek, and Thomas McCann, *Sophomore English Curriculum* (Unpublished curriculum guide, Community High School District 94, 1990).

Further Suggested Reading

Boostrom, Robert. *Developing Creative & Critical Thinking.* Lincolnwood Ill.: National Textbook Company, 1992.

Caine, Renate Nummela and Geoffrey Caine. *Making Connections: Teaching and the Human Brain.* Alexandria, Va.: Association for Supervision and Curriculum Development, 1991.

Gere, Anne Ruggles. "Introduction," in *Roots in the Sawdust: Writing to Learn across the Disciplines,* ed. Anne Ruggles Gere, pp. 1–8. Urbana, Ill.: National Council of Teachers of English, 1985.

Hatfield, Mary M. and Gary G. Bitter. "Communicating Mathematics." *Mathematics Teacher,* 84 (November 1991), pp. 615–621.

Johnson, Marvin L. "Writing in Mathematics Classes: A Valuable Tool for Learning." *Mathematics Teacher,* 76 (February 1983), pp.117–19.

Kulm, Gerald. *Assessing Higher Order Thinking in Mathematics.* Washington, D.C.: American Association for the Advancement of Science, 1990.

McIntosh, Margaret E. "No Time for Writing in Your Math Class?" *Mathematics Teacher,* 84 (September 1991), pp. 423–33.

Miller, L. Diane. "Writing to Learn Mathematics." *Mathematics Teacher,* 84 (October 1991), pp. 516–521.

National Council of Teachers of Mathematics, Commission on Standards for School Mathematics. *Curriculum and Evaluation Standards for School Mathematics.* Reston, Va.: The Council, 1989.

National Council of Teachers of Mathematics, Commission on Standards for School Mathematics. *Professional Standards for Teaching Mathematics.* Reston, Va.: The Council, 1991.

National Research Council. *Everybody Counts.* Washington, D.C.: National Academy Press, 1989.

National Research Council. *Reshaping School Mathematics.* Washington, D.C.: National Academy Press, 1990.

Pandey, Tej. *A Sampler of Mathematics Assessment.* Sacramento, Calif.: California Department of Education Office of State Printing, 1991.

Ruggiero, Vincent Ryan. *Teaching Thinking Across the Curriculum.* New York: Harper & Row Publishers, 1988.

Schwartz, Judah L. "The Intellectual Costs of Secrecy in Mathematics Assessment." In *Expanding Student Assessment,* ed. Vito Perrone, pp. 132–141. Alexandra, Va.: Association for Supervision and Curriculum Development, 1991.

Schmidt, Don. "Writing in Math Class," *Roots in the Sawdust: Writing to Learn across the Disciplines,* ed. Anne Ruggles Gere, pp. 104–16. Urbana, Ill.: National Council of Teachers of English, 1985.

Singer, Harry and Dan Donlan. *Reading and Learning from Text.* Hillsdale, New Jersey: Lawrence Erlbaum Associates, 1980.

Stenmark, Jean Kerr. *Assessment Alternatives in Mathematics.* Berkeley, Calif.: California Mathematics Council Campaign for Mathematics and EQUALS, 1989.

Vermont State Department of Education, The Mathematics Portfolio Committee. *Vermont Portfolio Program.* Montpelier, Vt.: Vermont State Department of Education, 1990.